Art and Culture

Big Ben

Shapes

Logan Avery

This is the Elizabeth Tower.

It is in London.

The bell in the
tower is called
Big Ben.

People call the
tower Big Ben, too.

Big Ben has circles.

There are circles
on the clock.

Big Ben has squares.

There are squares around the clock.

Big Ben has
a triangle.

There is a triangle
near the top.

Big Ben has
rectangles.

The side has
big rectangles.

Big Ben has shapes.

What shapes do you see?

Problem Solving

Draw your own tower.
Use one or more of
each shape.

- circle

- triangle

- square

- rectangle

- hexagon

Answer Key

Drawings will vary but should include one or more of each shape: circle, triangle, square, rectangle, and hexagon.

Consultants

Nicole Belasco, M.Ed.
Kindergarten Teacher, Colonial School District

Colleen Pollitt, M.A.Ed.
Math Support Teacher, Howard County Public Schools

Publishing Credits

Rachelle Cracchiolo, M.S.Ed., *Publisher*
Conni Medina, M.A.Ed., *Managing Editor*
Dona Herweck Rice, *Series Developer*
Emily R. Smith, M.A.Ed., *Series Developer*
Diana Kenney, M.A.Ed., NBCT, *Content Director*
June Kikuchi, *Content Director*
Véronique Bos, *Creative Director*
Robin Erickson, *Art Director*
Stacy Monsman, M.A., and Karen Malaska, M.Ed., *Editors*
Michelle Jovin, M.A., *Associate Editor*
Fabiola Sepulveda, *Graphic Designer*

Image Credits: p.4 Richard Bryant/Arcaid Images/Alamy; all other images iStock and/or Shutterstock.

Library of Congress Cataloging-in-Publication Data

Names: Avery, Logan, author.
Title: Art and culture. Big Ben / Logan Avery.
Description: Huntington Beach, CA : Teacher Created Materials, 2018. |
    Audience: K to Grade 3.
Identifiers: LCCN 2017059904 (print) | LCCN 2017061112 (ebook) | ISBN
    9781480759671 (e-book) | ISBN 9781425856298 (pbk.)
Subjects: LCSH: Shapes--Juvenile literature. | Big Ben (Tower
    clock)--Juvenile literature.
Classification: LCC QA445.5 (ebook) | LCC QA445.5 .A925 2018 (print) | DDC
    516/.15--dc23
LC record available at https://lccn.loc.gov/2017059904

Teacher Created Materials

5301 Oceanus Drive
Huntington Beach, CA 92649-1030
www.tcmpub.com

ISBN 978-1-4258-5629-8